Happy, Healthy & Hyperconnected:
Raise a Thoughtful Communicator in a Digital World

By Christa Melnyk Hines

I0170412

Happy, Healthy & Hyperconnected:
Raise a Thoughtful Communicator in a Digital World

by **Christa Melnyk Hines**

Dedication

To Nolan and Drew.
Two inspiring reasons why I wrote this book.

And to Jason.
For your love and support every step of the way.

Table of Contents

Introduction

Much of what we see screaming from news headlines is scary. Child predators, cyber-bullies, and reputation-ruining snakes are just waiting to phish out innocent victims from the safety of our homes through our wireless network. Keeping track of everything kids do, see, or say online seems overwhelming and just plain exhausting.

Then, there are the reports that the impulsive nature of our communication technology has created an alarming breakdown in the quality of our interactions. Researchers worry about the trend of isolation in our society due to a decline in compelling conversation and meaningful connection that contributes to community-building.

Now your child is asking—insisting—that she needs a smart phone. With all of the negative press about the impact of social media technology on our youth, it is no wonder parents are concerned.

The truth is we can't throw caution to the wind or take these reports lightly. This is information we need to hear in order to become more attentive about how our kids relate to others. The examples we see played out across the media reiterate the importance of guarding our children from becoming victims of something nefarious that will doggedly nip at their heels for years to come.

But as a mom, I get frustrated with these reports. Casting technology as the bad guy is misguided and unproductive. Our job is to better educate our kids on what *can* happen, not what *will* happen.

What if I told you that simply teaching your child how to communicate thoughtfully is one of the greatest lessons you can impart to address many of these problems?

When we make a conscientious effort to teach our kids to communicate well and considerately, we help them develop into confident, independent men and women who are conscious of the world beyond the latest device they hold in their hands.

These are the young people who will be best prepared to compete in our global world while using the resources available to them.

They will have a firm set of boundaries that help keep them safe.

They will be happier in their relationships because they will know how to relate to others and resolve differences amicably and respectfully.

They will be less likely to feel isolated because they will know how to use social media to connect in meaningful ways.

They will know how to speak with integrity.

Their connections with others will be stronger and healthier.

Technology can't teach our kids these skills. You are your child's best model and most trusted resource and ally when it comes to navigating the tricky world of offline and online communication.

As a communication researcher, a parenting journalist, a mom of two children, and a life-long observer of human nature, I'm convinced that a firm foundation in old-school communication skills and clearly defined boundaries can help our kids enjoy higher quality, more balanced relationships. They will feel happier with the individuals that they invite into their space, both online and off.

Recently, I watched a video in which a handful of parents enrolled their quiet, self-conscious tweens into an after-school public speaking program. Over the course of the program, the kids began to blossom in the supportive, encouraging environment. Many of the parents were astonished at how their previously insecure offspring began to carry themselves a little taller and speak more assertively in other areas of their life thanks to the presentation skills they had gained in the program.

These results don't surprise me. I've studied human communication for more than 20 years. Through my research and experiences in public speaking, social media technology, media relations and interpersonal relationships, I see one common thread. The people who are strong communicators have the ability to form meaningful, warm connections. They know how to relay their messages confidently and thoughtfully. They can distinguish between which mode of communication is necessary given the situation. And most of all, they are perceptive, astute listeners.

One of the biggest barriers to successful communication is uncertainty. With uncertainty comes anxiety. And, as human beings we typically work to avoid that which we fear.

In this book, I break down common scenarios that lead to uncertainty and fear. I show you how to teach your kids to become comfortable in multiple situations from face-to-face and on the phone to various online interactions. I help you address your fears.

You may assume some of these skills come naturally. But when you consider how much the communication landscape has changed, with multiple mediums competing for attention, you will quickly see how important it is to take a step back. Revived attention to traditional, old-fashioned communication values can better prepare our kids as they segue into the online world where a comprehensive foundation of skills will serve them well.

How to Use This Book

This book is divided into two sections, old-school communication skills every child needs to learn and new school communication skills every child needs to learn.

The first section covers everything from basic manners to training your kids to use the phone and navigating difficult and contentious conversations. These skills are a good starting point because they are the basic communication traits that will help you underscore the values you hold dear. These values humanize us, nurturing sensitivity and reinforcing the importance of treating others with dignity and respect.

Raised with a strong footing in old-school ways, your child will lean on these skills as he begins navigating the online world.

The second half of the book will explore the many ways in which our digital children are likely to communicate with others online. In this section, you will find information about how to text and use smart phones respectfully; and fun, enriching ways to play and learn in the online environment. On a more serious note, I will explore issues that concern many parents, like cyber safety, social media harassment and online reputation management.

In the appendix, you will find additional resources; information about guiding your kids to interact safely with people they don't know; as well as, a short section on social media addiction. I also point you to links on my website for current online resources and research.

Throughout this compact guide I boil down what you need to know now, no matter where you are on your parenting journey. You will find a realistic approach that features proactive and creative ways to help your child connect with the world in a healthy way. What's more, you will further strengthen your relationship as you work to build and maintain honest, open communication.

Best of all, you can begin integrating these simple ideas into your family's busy lifestyle right away.

This book offers a condensed look at nurturing communication skills in your family. It is a handbook that you may refer back to time and again as your kids enter different stages of development. I've found as a parent, and you will probably agree, being aware and proactive about what is out there before you have to deal with it is half the battle!

Finally, as a busy parent myself, I realize how valuable your time is. Like you, I seek resources that can deliver information quickly and help me address issues in my family through achievable, practical ways. I crafted this book with that goal in mind. I look forward to hearing your stories of success and encourage you to share your comments and thoughts.

Section I: Old-School Communication Skills Every Child Needs to Master

Without question, technology has changed the way we interact with one another. In many respects, our social networking capabilities have made life easier, creating circles of connection in profound ways. And we have the capacity to connect with more people than ever before.

While our society's love affair with social technology continues to evolve, the traditional ways of interacting with one another remain timeless. But many children confuse the various nuances that exist between offline and online communication. The rules we follow when we communicate on social media don't necessarily apply in an offline environment.

Old-school communication skills are still very much in vogue, indispensible and relevant in our communication-hungry landscape. Children raised with the ability to move between face-to-face, group, and phone interactions with polite, self-assured confidence are likelier to enjoy more satisfying interpersonal and professional relationships. Furthermore, these traits will serve them well as they learn to adjust to changing circumstances and to navigate their online conversations. In turn, your child will grow into an adult who is empowered and well-equipped to successfully pursue his dreams.

Chapter 1: Demonstrating In-Person Manners

Manners are a sensitive awareness of the feelings of others. If you have that awareness, you have good manners, no matter what fork you use.

~*Emily Post*

"Kids today. No manners." Every generation has grumbled about the next when it comes to basic etiquette. Past forms of etiquette are considered stuffy by today's standards, but for the most part manners come down to treating others with polite courtesy and respect. How do manners influence our social interactions, and how can we best teach them to our kids?

Why Manners Matter

Distinguishing us from a rabble of unwholesome cretins streaming out of middle earth, our society has developed a set of rules and expectations for polite social interactions. At their core, manners lay the foundation for a lifetime of successful communication skills. We rely on manners when navigating tough and awkward social situations, like when we must say no to a request that we can't or don't want to fulfill; while networking with people we don't know; or when we are working through a disagreement. From the mail carrier and grocery store clerk to family and friends, basic etiquette informs those around us that we are sensitive to them and aware of our surroundings.

Specific forms of etiquette are subjective and evolving based on your personal background, cultural norms and expectations. Nonetheless, helping your kids learn acceptable behavior across an array of situations while interacting with different generations will help them feel less anxious when entering into environments where they are uncertain about the norms. For example, when you go to an interview where you aren't sure of the work environment's expectations, you are more likely to act and dress in a more formal way. This shows respect to the people who are considering whether or not to hire you and helps you get a feel for the environment.

Uninformed, misguided behavior could ruffle the feathers of those around you, resulting in alienation from the group. This can negatively impact your ability to get your message across. What a shame if your important message, amazing idea or talent is lost on an audience who is too busy focusing on your lack of presentation skills.

The bottom line is manners form a framework that we can lean on to smoothly and politely navigate a wide array of social settings, whether we are amicably resolving a price dispute with a store manager, listening to a long-winded guest at a wedding or making small talk at a mixer. Furthermore, something as simple as "please" and "thank you" goes a long way toward nurturing your most important relationships, showing the people you care about that you love and respect them and don't take them for granted.

Role Modeling

Manners vary from family to family, with some parents preferring a more formal approach to others taking a decidedly relaxed view. Whether you require your kids to address adults as "ma'am" or "sir" or you have a more casual outlook and don't mind if your kids call your friends by their first names, most of us agree that basic etiquette is a necessary skill for kids to develop.

As a rule, kids look to their parents for guidance when it comes to how they relate to peers, family members, service industry employees and strangers. If you make eye contact, smile and make it a rule to say please and thank you during interactions with people, your children will learn that this is an appropriate way to treat others.

Holding the door open for a woman pushing a stroller or offering your seat on the bus for a pregnant woman or senior, for example, isn't simply old-fashioned chivalry. Such actions, whether you are a man or a woman, show people that you are a caring individual and that you are sensitive to others in your environment. Your kids will notice the warm response you receive when you carry out these actions.

Through role modeling and teaching them how to treat others, whether during a playdate or in a restaurant, kids grow increasingly cognizant that other people have feelings just like they do. These initial lessons in empathy help them understand why it is necessary to show courtesy and respect through both words and actions.

The seeds of social niceties that we plant now will guide them throughout their lives, including in the online world. And, they will develop an expectation of how they should be treated based on how they have learned to treat others.

Creating Boundaries

Etiquette norms not only set up first impressions, they establish boundaries between ourselves and people we don't know well or at all. These culture-driven parameters help us distinguish how to behave from one situation to the next. Chances are you don't behave the same way in a room full of clients as you do when you are at a party with your close friends. Similarly, kids learn that they have to use their "quiet voices" inside a classroom, reserving shouting and playing loudly for outside.

My grandmother jumps to mind when I think about social etiquette and boundaries. She believed strongly in treating others with politeness, but she had little patience for overly friendly strangers. She was suspicious of people who exhibited what she considered artificial friendliness. Although I personally like a warm approach, she had a point. We don't necessarily want to train our children to be robotic pleasers, but to respond to people courteously and politely with a firm set of boundaries in place.

In a world where people want things to happen faster and faster, including their relationships, manners are a gentle way to put our hand up and say, "I respect you as a person and look forward to getting to know you."

How to Teach Ps & Qs

Teaching manners takes patience and lots of repetition, which is why some parents choose to send their kids to experts for help. When I showed my sons an article that appeared in our local paper about a lady in our community who holds etiquette classes for youngsters, they both looked at me and said, "Mom, do NOT take us to that lady!" Did they note a gleam in my eye as I groused about their table manners?

While I liked seeing my kids carefully chewing their food with their mouths closed as they contemplated the horrifying prospect of etiquette school, I don't need to hire someone to teach them to eat more like civilized people and less like cave men. And you don't either.

From the time your little one can talk, integrate the words please and thank you into his vocabulary. For example:

"Gimme!" yells your child.

"Please," you say.

"Peas," your child repeats and you hand over the item.

As your child gets older:

"I want milk," he says.

"I would like some milk, please," you say and hold the milk as he says it back to you.

Around the dinner table, playdates, and whenever your child asks for something (which is probably a lot!), request that he ask with his "magic words" each time. Restaurants offer an especially rich opportunity for kids to politely practice ordering food and interacting with an adult they don't know. Jump in when your child struggles, isn't speaking up, or appears unsure how to answer a question.

Eventually, polite talk will become ingrained in your child's speech patterns. But, be patient. Your kids may see their peers not using the same set of manners that you emphasize, which can make your job a little more challenging.

Other skills your child should begin honing early on are a firm handshake, eye contact and confident posture. Kids love to play pretend and the next time your child invites you to participate in one of her games, teach her how to do the meet-and-greet.

Here's an example script for a meet-and-greet game:

> You: "Hello, Miss Katie. Thank you for inviting me for tea!"
>
> Reach out to shake her hand in a firm and welcoming way— no squeezes or pumping of the arm! You don't want to hurt her.
>
> If she looks at you funny and seems unsure what to do, whisper, "Okay, now you tell me, 'Thank you for coming today, Miss Smith. Won't you please sit down?'"

As your children get older, continue to help them practice a firm handshake, make eye contact and to smile when greeting people. Ask friends of the family or relatives to help. Use positive reinforcement to acknowledge the times she does a good job of greeting someone new.

Over time, even a self-conscious child will grow more confident about saying hello and interacting with adults. Soon, with greetings out of the way, she'll become more comfortable engaging in conversation.

Chapter 2: Making Conversation and Small Talk

Don't knock the weather; nine-tenths of the people wouldn't start a conversation if it didn't change once in awhile.

~ Kin Hubbard

Do you dread small talk? Some people claim that it is a meaningless waste of time. No matter what your attitude about it or how much you try to avoid it, small talk is part of life. Even though idle chit-chat may feel forced and sometimes even awkward, teaching your child to talk comfortably to people they don't know or only know casually is a useful skill. Social finesse can help them make new connections, handle the big events of life with poise and save them plenty of anxiety down the road.

Of course, I don't encourage my kids to talk to anyone and everyone. When I say "people they don't know," I mean this more in terms of interacting with store employees or new teachers. My kids know that if an adult they don't know well, or at all, approaches them and I am not there, they need to run away from that person. Adults don't ask kids for help. Tell them to run and get a trusted adult. They don't have to say a word. (Note: At the end of the book, I provide additional tips about how to help kids seek help from adults they don't know when you aren't present.)

The best way to begin teaching kids conversation skills is to talk to them when they are babies long before they begin to speak. Their brains are taking in and processing all the information you are sharing. For example:

- Describe what you see when you are out and about.
- Sing songs.

- Expose them to experiences to enrich their language skills, like going to the zoo or children's museum. Even a trip to the grocery store is a teaching opportunity.
- Read picture books together.
- Show them photos and discuss what is in the photos using descriptive language and complete sentences.
- Engage with your child directly and frequently. Television or talking in front of them to others won't nurture these language skills in the same way.

When it comes to small talk, you may wonder how you will teach this skill if you struggle with it yourself. The fact is you do know how. You socialize with your family all of the time. Put into practice, small talk presents a simple, but precious way to connect with your family that your children will rally around.

A Lesson in Dialogue

Here are a few ways to teach your children conversational skills:

Ask questions and listen. Start at the dinner table. Mealtime conversation is relaxed and helps your child learn the nuances of discussion. If the dinner hour is impossible, choose another time like breakfast, lunch or snack time.

Encourage your child to ask questions. I have a child who loves to talk to us, but sometimes he gets stuck on one particular topic that he is particularly passionate about. Children are naturally self-centered. We admire his passion and we support his interests. But I also don't want him to think it is okay to drone on about something that will bore other people to tears.

To help him learn to include people in conversation, I told him to ask each person at the table a question about something he knows interests them. He immediately turned to his brother and asked him, "what American hero would you choose to be?" (He did a project in social studies at school where the students each chose an American hero to research.)

He is a perceptive, intuitive child and I was proud of him for knowing right away what question would immediately engage his brother. We told him what a great question that was, and it started an entirely new conversation that the whole family embraced.

Headed to visit relatives or friends? Ask your children to come up with a few questions that they can ask. The ability to conduct themselves confidently will help them grow more assertive. Also suggest topics that they can bring up. "You know how you were asking if I was alive when the astronauts landed on the moon? Your grandparents watched it on TV. You should ask them about it."

Make Small Talk Fun

Want other fun ideas to encourage small talk at the table or even the car, and connect in a meaningful way with your kids at the same time?

- Chat Packs. Chat Packs are packs of cards that you can purchase that have conversation starters in them. When the conversation well is running dry or when my kids are goofing off at the table, this is a fun way to get their attention and reconnect.

- Play "what if." What if you could time travel? Where would you go? Why?

- Conversation jar. This idea is similar to Chat Packs except you write down your own conversation starters and put them in a mason jar. Each night, take turns pulling a topic from the jar. (Need ideas? Download free conversation starters on my website.)

- Draw. Spread butcher paper across the table and hand your kids some crayons or colored pencils. They can doodle and draw while you finish preparing dinner. Discuss their drawings during dinner. This activity appeals to kids of all ages.

- Current events. Have older kids? Ask each of them to come to the table ready to share a current event that they learned about in the news that day.

Even if you do not consider yourself adept at social skills, you can become more proficient together.

The number one piece of advice I can offer when it comes to engaging in social talk is this: Be curious. Be fascinated. Ask general (not-too-personal) questions. People usually love to talk about themselves and often they might say something that either sparks a connection or leads to another, more substantive conversation. As Deborah Tannen, professor of Linguistics at Georgetown University, says, "Each person's life is lived as a series of conversations." *

Examples:

"Where are you from originally?"

"How do you know the hostess?"

"I've been wanting to try that new Thai restaurant around the corner. Have you tried it yet?"

"Do you like comedies? Have you seen that new sitcom everyone is talking about?"

Conversation requires practice. But as you'll learn in the next chapter, the ability to listen is the secret weapon that sets the mediocre talker apart from the truly accomplished and dynamic conversationalist.

Quote reprinted with permission from Dr. Loren Ekroth, publisher of the newsletter "Better Conversations."

Chapter 3: Practicing Listening Skills

Good listeners, like precious gems, are to be treasured.
~ Walter Anderson

Listening is one of the hardest skills for any of us to learn and among the most respected. It requires focus and concentration. Think about that friend or family member who you can always turn to because they listen carefully to what you are saying.

Some of the best networkers I know can turn to a person and ask the question: "How are you?" and you immediately sense that they aren't just turning the phrase. They really want to know. You can feel their interest by their eye contact, their body gestures like nodding their heads and their follow-up questions.

Listening shows that you are focused and interested in what the other person has to say and you value what they say—even if you disagree.

When other people listen to us, we feel validated, like our opinions and thoughts have merit or at least matter to the person we share them with. This is no different for our children.

Promote Listening Skills

Here are a few games that nurture listening.

- Play Simon Says. An oldie, but a goodie... If you aren't familiar with this game, here is how it works. One person plays "Simon." He might say to the others, "Simon says, stand on one leg." Then, "Simon says, wave hello." Finally, "Now jump on one leg." The person who acts without the "Simon says" cue loses. By the way, this game comes in handy while stuck in an airport terminal with two stir-crazy preschoolers.

- Shout-out. Reading aloud to your kids nurtures listening and comprehension skills. Ask your child questions as you read the story. If the story has a repetitive word in it have your child shout it out every time she hears it. When my kids were younger, their personal favorite was *Five Little Monkeys Jumping on the Bed* by Eileen Christelow. My kids loved shouting out: "No more monkeys jumping on the bed!"

- 20 Questions for Kids. You can purchase this game or just make up your own version. Put slips of paper in a jar each with an animal, famous person or object. Each person pulls a slip of paper and the other players have to guess who or what the person is by asking questions. They have to listen to deduce the answer.

- Who am I? Gather your preschooler's stuffed animals. Have her turn her back while you use silly voices for each animal. Describe the animal and challenge her to figure out which one you are talking about. After she guesses, it's her turn to test you.

- Yoga. With names like gorilla, cat, flower, and airplane, many of yoga's poses are playful and appeal to kids' imaginations. Yoga is beneficial for kids for a number of reasons, including concentration, focus and listening. They must listen for the poses while moving their bodies. If you aren't familiar with yoga, but would like to try it with your kids, check out *The Kids' Yoga Deck: 50 Poses and Games* or look for kids' yoga instructions on YouTube. One well-respected teacher of kids yoga is Marsha Wenig.

- "I'm going on a trip..." My kids like playing this game in the car. The first person starts by saying: "I'm going on a trip. I'm packing toothpaste (or whatever item they choose)." The next person might say, "I'm going on a trip. I'm packing toothpaste and socks." Each player repeats the items already mentioned and adds her own. Don't be surprised if some of the items get a little outrageous! Players who forget an item are out. The last player who can remember the entire list wins.

When Listening is a Struggle

To further practice listening skills, help your kids distinguish between places where they can talk freely and those places where they will need to pay attention or pipe down out of courtesy and respect for others and to avoid missing out on important information. Go on a guided tour of a museum or rent headsets that require them to listen to a tour in order to understand the exhibits.

From the time they are little, ask that they listen when a speaker is sharing information like at the library, in a class or on the preschool fire department tour. Emphasize the need for calm quiet in places like the movie theater, church, the library, her sister's piano recital and concerts halls.

You can also create awareness that this is a time when listening is required with a non-verbal cue, like placing your hand on his shoulder and putting your finger to your lips.

If your child tends to frequently repeat himself to you, this may be an indication that he doesn't feel confident that you listened the first time around. Assure him that you heard him by saying something like, "Wow. You must have had a really fun time playing that game at Joey's. That's the third time you've told me that story!" Also, look at your child when he talks to you to temper his urge to repeat himself.

Become sensitive about how you communicate with others, especially when your child is present. Do you engage in conversation or a one-sided monologue when you typically talk to others? Do you frequently interrupt people when they speak?

When you are in the middle of a conversation or finishing correspondence and your child tries to interrupt, calmly excuse yourself momentarily from your conversation or pause to simply tell him to please wait so that you can give him your full attention when you are done with the conversation you are involved in. Later, explain to your child that unless there's an emergency, it's rude to interrupt. If it's an urgent matter, tell him to say "Excuse me..." Once he's old enough to understand, put your finger up as a sign that you are trying to listen to the person you are talking with in person or on the phone.

If you allow frequent interruptions without explaining that this is impolite behavior, you may inadvertently reinforce this behavior, and he will have a harder time learning the value of turn-taking in conversation and respecting people when they are engaged in conversations with others.

But, whenever possible, pay attention when your kids come to you and tell you a story. Put aside your email, turn down the radio, and ignore the ringing phone if you can. Look at your child, listen to her story, ask questions and be present. Today's seemingly unimportant stories that are dancing around their imaginations or the little dramas acted out on the playground will one day evolve into stories that really do need to reach your ears. Listen now, and they'll know you care enough to listen later. I promise, you'll never regret those moments.

Chapter 4: Learning Phone Etiquette

The telephone, which interrupts the most serious conversations and cuts short the most weighty observations, has a romance of its own.
~ Virginia Woolf

Phone conversation can create anxiety especially for people who prefer face-to-face interaction or online communication. Sometimes it is both simpler and a time-saver to pick up the phone and discuss a situation or misunderstanding, rather than continuing to swap emails or text messages. And even if you can't see the other person's face, you are able to discern the feeling behind the words through voice inflections and tone.

Consider this statement:

I don't want to go out tonight.
I *don't* want to go out tonight.
I don't *want* to go out tonight.
I don't want to go out *tonight.*

The emphasis on one word can change the meaning behind the entire sentence. Because the phone allows us to hear the person's tone of voice, we can politely and more accurately respond with the appropriate level of sensitivity the matter requires.

Teaching your child how to use the phone is both practical and necessary. Despite advances in communication technology, people still pick up the phone to conduct their business. Often it's simpler to speak to a live person, especially if you need a response right away.

Practice initially with a pretend phone or cell phone. Take advantage of your business travel to help your child practice phone skills. Call home to talk. Also, have your child initiate phone calls to grandparents who will be thrilled to hear their sweet voices. When no one answers, your child can learn to politely leave a message.

As your child gets older, have her phone a friend to invite her over to play. Since many families are converting to cell phones and moving away from home landlines, you may need to give the other parent a heads-up via text that your child is planning to call. Or at least be available to jump on the phone to provide support to your child and talk to the other parent.

While picking up the phone is relatively easy, using it in a way that is polite and respectful isn't necessarily a skill all people learn. But your child will.

Need a sample script?

> "Mrs. Jones [is Mrs. Jones there]? Hi, this is Becky. I was wondering if Danielle could come over and play at my house today."

Play the Telephone Game

Teach your child how to make an emergency phone call through role play. You pretend to be the emergency operator. This is also good way for your child to practice her address, a necessary skill going into kindergarten.

> You: "9-1-1. What's your emergency?"

> Your child: "My house is on fire. I need help."

> You: "Okay. Your name?

> Your child: "Vivian Jackson."

> You: "Vivian, what is your address?"

Your child: "3404 Sycamore Street."

You: "Thank you. Emergency crews are on their way. Are you out of the house?"

Your child: "Yes. I'm in my driveway."

You: "Is there anyone else in the house?"

Your child: "No, we all got out."

Explain to your child what to do if she reaches a wrong number.

The person reached: "You have the wrong number."

Your child: "Okay. I'm sorry. Thank you." is a sufficient answer before hanging up.

With practice and time, your child will grow more confident initiating phone conversations.

You can make these games entertaining by making up silly voices or names. The bottom line is you are teaching your child a fundamental skill about how to manage various phone situations. Anxiety is usually fueled from the fear of the unknown. Depending on their past experiences, your kids may not be able to articulate their worries, but may wonder: "Will this person I speak with treat me with dignity or brush me off as unimportant because I'm not a grown-up?"

Try these other scripts to help your child practice phone skills as he gets older:

How to make a doctor or dentist appointment.

You: "Good afternoon, Dr. Goodman's office, how may I help you?"

Your child: "Hi, I'd like to make an appointment for a physical."

You: "Your name and date of birth, please..."

Your child: "My name is Jack Leigh. My birthday is December 12, 2005."

You: "Thank you, Mr. Leigh....I can get you in on December 8 at 1. Will that work for you?"

Your child: "Could I make the appointment over Christmas break? The dates are December 18 through January 9."

You: "Yes, we can do that. How about December 19 at 3?"

Your child: "Yes, that will work. Thank you!"

Make an appointment for a haircut.

You: "Good morning, Chopshop."

Your child: "Hi, I'd like to make an appointment for a haircut."

You: "Any particular stylist?"

Your child: "Is Stella available?"

You: "We can get you in with Stella on August 9 at 4. Does that work for you?"

Your child: "Yes, thank you."

You: "Can I get your name and phone number, please?"

Your child: "Janna Smith, 818-999-2424."

You: "Thank you! We'll see you next week!"

Your child: "Okay. Thank you!"

How to order a pizza.

You: "Pizza Joe's."

Your child: "Hi. I'd like to place an order for delivery, please."

You: "Name and number?"

Your child: "Danica, 743-123-4466"

You: "What can I get you?"

Your child: "We'd like a large pepperoni, please. And, I have a coupon."

You: "Thank you. Anything else?"

Your child: "A large salad and some breadsticks, too, please."

You: "Thank you. I've got a large pepperoni, large salad and an order of breadsticks. Anything else?"

Your child: "That's all. Thanks!"

You: "Thank you. That will be $25.99. Will this be cash or charge?"

Your child: "We'll pay cash."

You: "Thank you. We should be there in 30 minutes."

Your child: "Thanks!"

Although most people rely on caller-ID to distinguish between telemarketer calls and friends, it's a good idea to teach your child how to answer the phone if he chooses to do so when you aren't there to pick it up.

How to answer the phone:

Your child: "Hello."

You: "Hi, is Janet there?"

Your child: "She can't come to the phone right now. May I ask who is calling?" [Show your child where the pen and paper is in case the caller wants her to take a message.]

You: "When will she be back?"

Your child: "I'm not sure when she can talk."

Caller: "Okay, I'll call back later."

You child: "Okay, good-bye."

In this way, your child is remaining respectful even though the other person is coming across as rather demanding and refuses to identify himself. She is managing her boundaries by not allowing the caller to intimidate her, and she is not providing personal information to someone she doesn't know.

Since some adults aren't always as patient with kids, treating them as if they are wasting their time, these are ways you can help your kids prepare for the unknown and speak with confidence. In the beginning, be there to jump in to provide reinforcement and support. As your child becomes accustomed to dealing with various individuals and expectations, her confidence will blossom, too, and people will respond with courtesy and respect.

Chapter 5: Expressing Gratitude

Acknowledging the good that you already have in your life is the foundation for all abundance.

~ Eckhart Tolle

Gratitude is one trait that never goes out of fashion. Pick up an Ann Landers, Miss Manners or any other advice column, and you will likely find at least one letter from someone feeling hurt that a gift or service went unacknowledged.

Whenever my children receive gifts or invitations for play dates, we have a family rule that an acknowledgement is owed the gift-giver or the hostess. Some people argue that if they didn't like the gift, they shouldn't have to thank the gift-giver. But, gratitude isn't so much about appreciating the object, service or time that was given; it is more about acknowledging the thoughtfulness and generosity of the giver.

Since my kids have been able to speak or hold a crayon, I have asked that they show thanks. From pictures to homemade cards or a phone call, they learn to express their thanks and in the process acknowledge the people and the gifts that come into their lives.

Practicing gratitude helps us reflect on the goodness that shines in our lives, and it is a healthy spiritual practice.

We all have bad days. You get a flat tire while it's pouring down rain. You spill your coffee all over the computer. (Gulp.) You have an upsetting disagreement with your spouse. You and your kids are miserably sick and your spouse is out of town. Open the paper or flip on the news and you probably won't feel much better. We can't avoid life's arrows and suffering. But when we focus too much on the negative, we risk growing increasingly cynical. Over time, it can become harder to see each day's gifts, and more difficult to find the silver linings.

Enter gratitude journals. When I've had a bad day, I write down three things that I'm grateful for. This helps me change my perspective even if those three things aren't particularly earth-shattering. On a particularly rotten day, I've had to scrape the barrel. "The sun was out. I found a penny. D took a nap." Seeing even the smallest blessings helps drown out the sorrows.

I try to share this attitude with my children. At dinner time, we occasionally share three things we are each grateful for that day, especially when we're all feeling a little down in the dumps or put upon. Bedtime is also a good time to reflect on the day when your child is troubled or feeling anxious and worried.

Expressions of gratitude are important in other areas of our life as well. Children who have developed a practice of thanking individuals learn that doing so helps positively nurture both personal and professional relationships.

The ability to write a thank you note is an asset and even though it is often recommended, it is becoming less common. Eventually, when your kids go to apply for a scholarship, an internship or a job where they are interviewed, crafting and mailing a professional thank you note will help distinguish them from other applicants. And since you've already been emphasizing this skill from the time they were young, it will be easier for them, too.

Accepting Compliments

Compliments are gifts of words and when they are genuine, they can be more meaningful to the recipient than an actual present. There's nothing like an honest acknowledgment of hard work and tenacity to make a person's heart do a happy little flip. But like adults, kids are often unsure how to respond to a compliment. And, that is probably because we adults aren't always the best role models in this area. Kids see us shake off or downplay these well-intentioned remarks.

Many of us become self-conscious or feel the need to temper a compliment with a response like, "Oh, it needed more salt." or "Thanks. All the other dresses I tried on made me look fat." Part of this comes from being brought up in a culture that values a sense of humility. We don't want to appear arrogant. However, the ability to graciously accept a compliment is an important communication skill and will help your child grow into an adult who is comfortable both offering and politely accepting compliments. Think of it this way: by knowing how to acknowledge and accept a compliment in a good-natured way, you honor the person who offered it.

We can accomplish a graceful response while maintaining our humble attitude by simply saying, "Thank you! I appreciate that!" or "Thanks! I'm glad you liked it!"

Teach your child to politely acknowledge a compliment even if she feels bashful about it or doesn't believe it herself. Tell your kids that all they have to say is thank you.

When an adult tells my son, "Hey, great game out there today. You played awesome defense," I've noticed that my six-year-old instinctively wants to look down and pretend he didn't hear the compliment. This is partly due to feeling self-conscious and also being unsure how to respond. I find myself nudging him and whispering, "Look at him and say thank you." With practice, kids grow more confident in these situations.

Offering Compliments

As long as you aren't fake (people can almost always sense if you are being disingenuous), giving compliments is a nice ice-breaker for making friends. If your child struggles with connecting and making friends, suggest that she say something kind to a peer. For example,

"I really like those shoes. That's cool how you added sparkly gems on them."

Putting Gratitude into Practice

Here are a few ways to help your child learn to practice gratitude, as well as gracefully give and receive compliments:

- Connect at dinnertime. Go around the table and ask each person to share the ups and downs from the day. Start with the negative, but always commit to ending on a high note with the positive. Even little things like a pretty sunrise count!

- Swap compliments. "Don't you look nice, Lauren!" "Thanks! You do, too, Mommy!" "Thank you!" Now, didn't that feel good?

- Make gratitude festive. After a birthday or holiday, have a family thank you note party. Craft cards and write notes together while listening to your family's favorite music. If your child is too young to write her own notes, write the notes for her, but have her draw or paint a picture, place stickers on the card, include her thumbprint or sign her name.

- Snap a picture. Take a photo of your child holding the gift and turn it into a thank-you note that she signs.

- Go digital. Video your child holding the gift and thanking the sender. Email the clip as a thank-you card.

- Create a blessings jar. Keep a stack of colorful notepaper and a pen next to a mason jar. Invite your family to write down the names of people or events and gifts that they are thankful for. Weekly or monthly, make a festive dinner or dessert and go through the jar together, acknowledging the special gifts and remarkable people in your family's life.

- Bake a treat. If someone went out of his or her way to help your family, consider delivering a sweet treat to that person. Invite your kids to help you put a small basket together. Even if they don't get involved in the process, they'll pay attention to what you are doing and why you are doing it.

The ability to show gratitude helps us combat cynical thoughts that can trample our personal happiness. Our children begin to see that outside of their inner world, people are generally kind, thoughtful and helpful. What a powerful, yet simple way for our kids to appreciate these qualities in others and begin to exhibit these conscientious traits themselves.

Chapter 6: Making Mistakes and Apologizing For Them

Never put off repairing a relationship you value. If sorry needs to be said, say it now. Tomorrow isn't guaranteed to any of us.
~ Toni Sorenson

Accepting accountability for our mistakes in how we communicate with others takes courage. But when we are willing to bow our heads and apologize, we strengthen our relationships, build trust and show integrity.

How we manage our own emotional mistakes can influence how our children will manage theirs. When you lose your temper or say something hurtful, apologize. Explain how you could have handled the situation better.

"I'm sorry I lost control of my temper today. I'm not proud of the way I handled myself and in my anger, I said things that I wish I could take back. I hope you will accept my apology and forgive me."

What doesn't matter is who is right and who is wrong. People will withhold apologies because they feel that they don't need to apologize when they believe they were right, even if they proved their point in a way that was harmful to the relationship.

In kid language that might look like this example:

> "This toy belongs to me so I can take it from you whenever I feel like it."

> Child grabs the toy away to the surprise of the other child. The other child cries.

The parent says, "Jeremy, you say you are sorry to Alex for grabbing the toy."

But, Jeremy doesn't understand why he should say he is sorry. "It's my toy!" he screams.

Help your child understand by asking him how he would have felt if his friend Alex had treated him like that. Explain that instead of grabbing the toy, he should have asked for it back nicely. It's not about the toy. It's about the suffering we caused to our friend's spirit. As my son used to say, "Mommy, my heart hurts."

Asking for Forgiveness

Asking for forgiveness seems so basic, but apologies are often mistaken for a sign of weakness. The truth is, you can stand on your hill of self-righteous indignation as long as you like, but hanging out with your opinion rather than the people you care about can get lonely.

By expressing an apology, we effectively tell our loved ones that we understand that how we treated them caused them suffering and we feel badly about it. A sincere apology works to rebuild trust in a relationship and it exhibits empathy.

Will trust be restored right away? It depends on the gravity or nature of the mistake. Sometimes we have to work a little harder at rectifying the damage that our mistakes and words have caused to other people. Asking for their forgiveness is a good first step.

Three ways to help your child say "I'm sorry.":

- Pull your friend or sibling aside and say, "I am sorry for how angry I acted. I should have handled it better. I hope you will forgive me."

- Write a letter. "I'm sorry for breaking your favorite toy. I will replace it. I hope you will forgive me."

- Offer a token. "I drew you this picture to show you how sorry I am for hurting your feelings. I hope we can still be friends."

How not to say sorry:

- "I'm sorry, but..." It negates the intention and turns the apology into an excuse.

- "Well, SOR-R-R-Y! Sheesh!" You are only saying it because you feel like you have to.

- "Oh. Sorry." Saying sorry can become an automatic, addictive response and is used as a
quick way to ease tensions. It usually doesn't mean anything.

When Apologies Aren't Accepted

Not all apologies will be accepted right off the bat. Tell your child that the best she can do is offer her authentic apologies and hope for the best outcome. Explain that it may take time for her friend to come around and to rebuild any trust lost.

In the best-case scenario, the friend or loved one gracefully accepts the apology. You'll hug or shake hands and move on. But people don't always do that and sometimes they just aren't ready to offer forgiveness. Here are a few not-so-great ways people respond to apologies:

- Gloating. "That's right! So you admit that I was right and you were wrong." Your friend is not admitting that. He is simply saying that he is sorry for his behavior.

43

- Making the friend feel worse. "You should be sorry. I can't talk to you or be your friend right now."

- Use it as power over someone. "If you are truly sorry then you'll prove it by...."

In these cases, it's best for your child to say to herself: "Well, I did my best to make up for my mistake. That's the best I can offer right now. If she won't accept my apology then I'll just have to move on."

Sometimes, time to heal is the only viable option and other times, your child learns a tough lesson in loss of a friendship. While she may feel wretched about the screw-up, help her see that she needs to forgive herself, too. We're only human after all.

Chapter 7: Resolving Conflict

We must teach our children to resolve their conflicts with words, not weapons.

~ William J. Clinton

Like every human, kids have conflict. The successful communicator experiences fewer conflicts because she knows how to manage these types of interactions assertively and proactively.

For kids who tend to be pleasers, they may experience more internal conflict because they are unsure how to say no to a request, activity or situation that causes them a sense of unease or dread. They may acquiesce to whatever the other person wants because it's easier than fighting about it. Fighting causes them too much stress.

Instead of stating their desires, pleasers might respond in a more back-handed way, such as by simply not following through on the promise they made and avoiding accountability by blaming someone else; gossiping and complaining about the other person behind their back; or suffering through it because they don't know how else to get out of the situation. How many of us over-promisers can relate to this problem?

Children who are used to getting their way or who rarely experience denial of their wishes and demands, may not know how to manage relationships where their equally sure-minded friend pushes back. Out of frustration, they may resort to name-calling, threats, tattling or physical aggression.

"Janie said she won't be my friend unless I play the game that she wants."

"Jackson said I was dumb and he took the ball away from me."

"I hit Jordan because she wouldn't give me the doll that I wanted to play with. It's mine."

"David won't quit touching my stuff in my room. I don't want him in my room."

Children can learn to respectfully resolve many of their conflicts with a little guidance from you. Beginning from the time they are toddlers, encourage sharing and take advantage of empathy-building opportunities, with scripts such as:

> "Wouldn't you feel really sad if Johnny had crushed your Lego tower? How can you make it up to him? How about you say you are sorry and offer to help him rebuild it?"

Model Assertiveness

Individuals who are described as assertive behave confidently and state their needs clearly, respectfully and directly. An aggressive person, on the other hand, attempts to force others to his will either through bullying or passive manipulation.

To help your child learn what it is like to stand up for himself in a non-aggressive way, model assertive, respectful behavior in front of him. For example: "We'd prefer not to dine next to the bathroom. Is there another place we could sit?" or "I was actually next in line. Thanks!"

Assume that people generally have good intentions. Customer service is hard, often thankless work. State your needs or concerns calmly without being a bully yourself. A common mistake is aggressively over-reacting to a benign situation and using statements like "always" or "never." This tactic rarely gets the results you are looking for—and what are the results you are hoping for, anyway?

Kids do this frequently when they complain. "He *always* gets his way. I *never* get what I want."
They say this out of frustration and anger without really considering the "always" and the "never."

When you find yourself wanting to behave similarly in public, consider how you could approach the situation in a way that models resolution and achieves proactive results rather than fuels negative reaction and defensiveness. For instance, you may want to announce angrily: "You always get our order wrong." "You people never have your act together when we come here."

While this may seem to be historically accurate, ask to speak with a manager about your concerns and use specific examples to make your case. Your kids will see you calmly address the situation. You might say to the manager: "We love to come here and eat. Your restaurant offers a great atmosphere and friendly servers. But, I thought you might like to be aware that the last three times we've come here, the food has arrived cold."

In turn, tell people in front of your kids when you appreciate the work they are doing, particularly if they've gone above and beyond your normal expectations. "Thank you for taking the extra time to help us search for the types of cleats we were looking for. I really appreciate your assistance."

And, positively reinforce your child's conflict management skills. When your child problem solves an issue between herself and another child, whether a friend or sibling, praise her. For example, "Good job coming up with a solution that works for both of you!" or "I like how you guys are taking turns with the video game. That was a great idea!"

Managing Aggression

Encourage your kids to use their words rather than their fists. Help them understand where the other person is coming from to build empathy. They can learn to treat each other with respect even if they dislike what the other person is doing or saying.

Ignoring, walking away or disengaging from another child who is teasing or making fun of your child is sage advice. Arming your child with a few good responses for his communication arsenal is also helpful. Your child will feel more empowered to stand up for himself, and he will increase his self-confidence in the process.

Without getting emotional or retaliating in a way that could get your child in trouble, he is in essence saying: "I don't allow people to talk to me/treat me like this."

When my son was in second grade, he often got reminders from his teacher to pay attention and listen. Another classmate, who had been a problem for my son for months, picked up on this and would mock him by repeatedly saying, "Are you listening?" It was really starting to bug my son, and he didn't know what to do about it.

I role-played some ideas that he could say in response like, "What?" in an unemotional tone every time the kid teased him with this question. I coached him by pretending to be the kid who teased him. The role-playing helped him practice saying what he wanted to say in just the right tone. No whininess or weak-sounding inflections allowed. He quickly had his chance to try it, and the obnoxious comments stopped.

Teachers and adults can't always be there to intervene in these situations and when a child has the skills to stick up for himself, he feels more empowered. Nonetheless, reassure your child that you and his teachers are there to help whenever he needs it. He does not have to manage bullying behavior alone.

Dr. Laura Markham offers some great tips on her blog Aha! Parenting.com to help "bully-proof" your child.

Here are some sample responses to coach your child to say if she has a hard time standing up for herself:

"It is my turn now."

"Don't touch me."

"I don't like being called by that name. Call me by my name."

"No, thank you."

"I want you to stop teasing me. What you said is mean."

Know When to Walk Away

Physical kids push each other around, jostle in lines and shove each other. Other kids use verbal jabs to intimidate, get what they want or control a situation. There will always be kids, just like adults, who cheat, say mean things and cut in line. Even if your child stands up for himself, the other kid might simply blow him off and push him aside anyway. Explain to your kids that we can't change other people's behaviors. We are only responsible for our reactions. And sometimes the only response that makes sense is to walk away and not take it personally. Instead, we might say to ourselves, "Jeez, what's that guy's problem?!"

If the aggressive individual continues to cause problems and invade your child's space, encourage your child to seek guidance from a trusted adult. He might say, "I told Josh that it was my turn, but he still pushed ahead of me. What should I do because he does this to me a lot?"

As children mature, the nature of their disagreements will change. Teaching them how to disengage and redirect a conversation is the best option when the timing is bad or when they don't feel emotionally prepared to have the argument.

Remember, you are not teaching your child to be passive or non-confrontational. Rather, you are helping her diffuse a situation in which she feels she lacks the ability to control her temper. We all have moments where we just don't feel like talking about a touchy subject, especially if we run the risk of making a public spectacle of ourselves or saying something thoughtless that we'll regret. Tell her she can always return to a disagreement when she feels a little more detached emotionally.

From an intellectual point of view, the best-case scenario is to approach a difference of opinion with the intent to learn rather than to change a mind. Many adults struggle with managing their emotions during tense conversations. Expecting a child to keep her cool in similar circumstances may be a lot to ask. Provide your kids with a few exit strategies to disengage from an uncomfortable conversation gracefully.

Consider the following example and three different ways your child can dissipate the tension:

Child's friend: "Why did you have to go and choose the dumb desert project for our science group? Are you crazy? It's going to be way harder. I wanted the ocean project. This is going to be so boring."

Another friend jumps in agreeing. Your child feels the need to defend herself and not the issue.

Response: "I don't like being called crazy. I had to make a decision and you guys weren't there. Clearly we don't see eye-to-eye, but I bet we can find a way to make this work." In this way, your child is diffusing the situation by keeping the discussion factual while standing up for herself.

Your child starts to feel overly emotional.

Response: "I'm not in the mood to talk about this right now. I'm getting angry and I don't want to say something mean." Or "Thanks for letting me know what you think. I guess you'll just have to have your opinion, and I'll have mine. Let's talk about it some more later."

Poor timing, like if the friend brings it up at a birthday party.

Response: "Wow. There's a lot I'd like to say about that, but this isn't a good time. I'll call you tonight and we'll talk. Hey, there's Jackie. Let's go say hi!"

Dealing with Sibling Rivalry

What do you do when home is the central battlefield? Your children may have been brought up under the same roof with the same set of rules, but none of that means anything when it comes to managing their personality differences and moods.

> "Will you stop crunching your cereal so loudly? It's getting on my nerves."

> "Get OUT of my room!"

> "I didn't tell you that you could wear my new sweater—and you ripped it! M-O-O-OM!"

> "It's my turn to play the video game. Give it to me now!"

While sibling rivalry is an age-old problem, that doesn't make the behavior any less frustrating for parents. Here are few ideas to help if conflicts are creating daily friction in your household.

- Create a resolution jar. Your children must work together to earn tokens. Using cotton balls, poker chips or marbles, award them with a token every time you catch them problem solving, sharing or teaming up together. When the jar is full, they get to choose an award like going out to dinner, watching a movie or planning a family game night—whatever incentive works well for your family.

- Play Rock, Paper, Scissors. Given the opportunity, many kids find that playing this game is a perfectly acceptable way to fairly solve a small disagreement, such as who gets to go first in a game, who gets to play with a particular toy first and so on. Remember, rock crushes scissors, scissors cut paper and paper covers rock. If both players choose the same, they must go again.

- Five minutes and swap. When my kids were preschoolers and had trouble sharing, this tactic worked well: "Okay, Tommy, you get to play with the car for five minutes. I'll set the timer. When the timer goes off, you have to give the car to Joey. Then he gets it for five minutes." Usually after 10 minutes, sometimes less, both kids had lost interest and moved onto something else. The conflict, it turns out, was more a power struggle than a desire for the toy.

- Retreat. Encourage your kids to take time to cool off when they are angry. Suggest they head up to their own rooms to color, listen to music, or read a book. Or ask them what helps them cool off. They might tell you that they need to go shoot hoops in the driveway or go for a bike ride.

- Discuss it. Ask your kids what they think would be a fair way to resolve a point of contention and offer some ideas if they aren't sure. I always think of a peanut butter commercial that ran awhile back where there was only enough sandwich bread for one peanut butter sandwich. Two brothers were fighting over who got to have it. The mother told one of her sons to cut the peanut butter sandwich in half and the other got to choose which half he wanted. Although the mother in this example helped resolve the issue, she did it in a way that both boys agreed was fair.

Learning to resolve conflicts without drama isn't easy for any of us, especially in the heat of the moment. But when we give ourselves permission to take time to collect ourselves before proceeding with a disagreement, we can head off a situation before it devolves into finger-pointing, hitting, name-calling and door slamming.

Section II: New-School Communication Skills Every Child Needs to Master

Treating others with dignity and respect shouldn't change when we head online into social networks or when we pick up our smart phones. But for some reason, these technologies have flipped common courtesy on its head, cutting into face-to-face interaction much to the annoyance of the people who aren't similarly engaged. In forums where people conduct themselves anonymously, some feel free to road rage through a thread. Additionally, distracted drivers are more focused on incoming texts and phone calls than they are on the roads creating dangerous conditions and spurring legal action.

As parents today we have the distinct advantage of curbing some of these issues among our developing "digital natives" through communication, education, and awareness. You don't have to be tech savvy to raise a communicator who knows how to treat others well both online and off. While I'll cover cyber safety and reputation management to a certain extent, this next section will provide you with ideas about how you can integrate technology into your child's life in a proactive, positive way while emphasizing the values that you want your child to live by.

(*Note:* Because popular apps and social media platforms change quickly among adolescents, I include an evolving list of sites to have on your radar on my website, www.christamelnykhines.com.)

Chapter 8: Teach "Cyber" Street Smarts

Safety is something that happens between your ears, not something you hold in your hands.

~ Jeff Cooper

As a parent, you probably set parameters for your child when you send her outside to play. Before she goes, you remind her to look both ways before crossing the street, to never approach a car and to always let you know where she is going. Basically, you give her the knowledge to keep herself safe—not to scare her.

With all of the expert warnings, nefarious cyber-criminals, cruel cyber-bullies and reports of social media behavior gone bad, many parents would just as soon shut the door on Internet access altogether. But that isn't realistic. We have to talk to our kids about how they can protect themselves, create and respect boundaries, communicate in a positive way and make good choices.

Nothing covert going on here!

While it is wise to be practical and put security and lock-down measures in place, education helps eliminate the mystery and curbs a sense of intrigue that compels kids to behave secretly.

Familiarize your child with the web and its many capabilities. My son used to be enamored with wind turbines. We'd explore the science and engineering of modern windmills and watch YouTube videos of workers constructing the massive structures. My other son and I spent several mornings watching National Geographic video clips from Shark Week during a time when he couldn't soak in enough information about these fearsome predators. If your child asks you a question and you aren't sure of the answer, take them to a search engine and together find out the answer.

Role model how you use the web as an information hub for news and subjects that interest you. Also, show them your social media pages so they can see how you connect with long-distance relatives and friends. Provide them with examples of spam messages that you see and how you know that the links, if clicked, could create havoc on your computer and put your information at risk.

If you blog, show your kids how it works and how you interact with people, including individuals who comment in disagreement with something you posted.

See something pop up on the page that you wish he hadn't seen? Explain that just as you have to separate the trash from the treasure on television, the web is a space that is as populated with valuable resources as it is with garbage. Tell your kids that your policy is to ignore weird pop-ups and provocative ads that can cause problems on your computer. And help them learn to vet information. My kids were under the impression for a while that "if it's on the Internet it must be true."

Use opportunities that you encounter on television or in the news to discuss examples of good and poor decision making when it comes to Internet use. Your child will probably be much more receptive about talking through another person's errors than listening to a lecture about what not to do.

Empower your child with information and a savvy awareness to navigate the cyber highway in a thoughtful way. Encourage her to lean on her sense of right and wrong that you've nurtured from the very beginning and your child can be just as safe surfing the net as she is playing outside with friends.

Informed about what's what and how to manage it, you and your kids will feel more prepared for a host of situations—and the ones she's not sure about, she'll know she can come to you to help problem solve.

Protecting Their Identity

We all know that our digital footprint holds valuable personal information, but you may not realize that your child's information is at risk, too, even if she isn't on the Internet yet. Because of a minor's clean credit history, identity thieves like opening credit cards in a child's name knowing that the theft can go undetected for years. Not only are they vulnerable to identity theft, information in the wrong hands can threaten a child's reputation and physical safety.

What measures should you take to protect your child?

- Other than for tax purposes or your child's school and medical records, secure your child's social security number under lock and key. Don't carry it in your purse or leave it laying around.

- Push back on organizations that ask for your child's SSN. Most don't really need that information to do business with you.

- Monitor how much information you post about your child on company websites. If the company's customer information is hacked, a child's birthday, age and place of birth are good starting points for thieves.

- Avoid posting your child's birthdate, age and place of birth online or in a baby gift registry. Make generic online birth announcements and ask the company to remove your child's gift registry after you are done with it.

- Each year, run a free report on your child through one of the three credit reporting agencies, including Experian, Equifax and Transunion. If a report shows up, there's a strong chance their identity has been used fraudulently.

Contact the creditor immediately to freeze the line of credit. You'll need to provide proof of guardianship.

- Keep your computer secure if you use it to store social security numbers or tax information.

Source: www.IKeepSafe.org

Create a Passport to Travel

A digital citizenship contract is one way to establish a clear set of rules, expectations and boundaries when it comes to your family's technology use. For example, you might make a list and discuss each item on the list, answering any questions your child might have. Make it an official looking document so your child understands the seriousness of your expectations. Agree on consequences if the contract is broken, such as the removal of computer or smart phone privileges for a designated period of time.

Here are a few ideas to add to your contract:

- I will use kind words while online and refrain from profanity.
- My actions towards others will be kind and respectful.
- I will treat others as I wish to be treated.
- I will not use my technology to bully or spread gossip.
- If I am bullied or attacked online, I will let mom or dad know right away.
- If I come across any information or photos online or receive a message that makes me uncomfortable, I will let mom or dad know.
- I am not responsible for other people's online behavior. I will not respond to messages that are abusive, gossipy, cruel or hurtful.
- I will not send photos of myself that could reflect poorly on my character or values.

- I will not forward photos or videos that could cause another person harm.
- I will not be in trouble if I realize I made a mistake online and notify mom or dad right away.
- I will not click on unknown links from people I don't know.
- I will ignore people who I don't know who try to connect with me online. If they continue to bother me, I will let mom or dad know and block that individual.
- I will steer clear of chat rooms in which participants are anonymous and/or where participants are verbally aggressive.
- I will not "friend" people I don't know.
- Before downloading apps to my computer or phone, I will talk to my parents first.

A digital citizenship contract gives you a prime opportunity to discuss different issues that can occur online and how to use online media in a healthy and proactive way.

Also, maintain access to all of your child's passwords and stay tuned to what he is downloading on to his phone and computer. Some parents set up a family iTunes account to keep track of the apps their children are downloading. Notice anything questionable? Instead of simply saying no to an app, explain why it doesn't fit with your family's technology-use goals.

Parent "Spies," Protection and Privacy

You absolutely have the right to go through your children's text messages and social networks to check up on them. If you are worried that you are intruding on your child's privacy, think about how being proactive now will save you from being reactive later. Catching a small mistake in judgment can mitigate any long-lasting repercussions. Unlike a personal diary tucked under the mattress, anything posted online is there for all to see, is easily shared and never goes away.

Some kids think it's no big deal to share their passwords with their best friends or their current flame. Explain that while it is okay to assume that your friends would never hurt you, the best policy is to protect yourself anyway. One way to help your child understand this concept is to tell her that just like you would never give your ATM bank card PIN to anyone, her online reputation is like social currency that needs to be protected.

Before sharing photos or texts, advise your child to always ask these questions:

- Is there any way at all, I could regret this later?
- Is this a text or photo I would want my mom or dad or a college admissions person to see?
- Would I be embarrassed if the whole school saw this?

What Not to Share

While I recommend that kids should stay off sites in which participants are anonymous, if they do comment in a public forum they should avoid sharing too much personal information since they don't know exactly who is lurking or how the information could be used. You should never share your:

- Social Security Number.
- Address.
- First and last name. (Use first name only or a pseudonym.)
- Birth date with year.
- Photos that identify where you are.
- Name of your school.
- Specific town where you live.

Explain that any time we express our opinions online in a public forum like a news site, others may respond with ridicule, anger or hate. Praise her for bravely sharing her thoughts and encourage her to consider the constructive criticisms other commenters post, while ignoring inflammatory comments, usually posted by so-called Internet trolls or negative people with little else to do with their time.

With a common-sense approach, you will guide your child through cyberspace just as you do in her daily offline interactions. And thanks to your thoughtful efforts, the online world will be one citizen stronger who knows how to responsibly manage online interactions with confidence and integrity.

Chapter 9: Coach Basic Smart Phone Etiquette

Approximately 80% of my regrets involve hitting send.

~ posted by a teen on her Tumblr site

When I am seeking a babysitter, I generally text because they are more responsive than if I call and leave a voice message. I don't mind using the mode of communication that they prefer.

That said, adolescents don't always seem to understand that while texting is a little less formal, they should approach it differently when communicating with an adult who is interested in employing them versus a peer or their parent. Remind your kids to remain a little more formal texting with an adult who they work for or don't know well. Reserve text lingo for friends. Write in complete sentences and show cordial respect in the same way she would do in person.

Consider this text interchange:

> You: Hi Cornelia. Would you be available to babysit for us this weekend?

> Cornelia: no, I cant. sory

Now compare that interaction to this exchange:

> You: Hi Janet. Would you be available to babysit for us this weekend?

> Janet: I'm sorry, but I have a volleyball tournament this weekend. I'd love to another time though!

Which teen are you most likely to go back to and ask to watch your little ones?

In the first response from Cornelia, here's what is going through my head as a mom: "She must not like to babysit for us...Did my kids give her problems last time? Hmm. Maybe she doesn't need the money. Maybe she is working now? I don't know. Oh well."

In reality, Cornelia may wonder why you never call her anymore.

The truth is, they don't even have to tell you why they can't babysit. A simple — "No, I'm sorry I can't this weekend. I'd love to another time." —would suffice, too.

When to Text, When to Talk

Evaluating different situations to decide on the mode of communication is also a skill every child should learn.

Young kids can practice their writing skills when they text mom or dad.

"Hi, Dad! I miss you! Love, Tyler"

When my kids are outside playing with friends and then go into one of their friend's houses, they'll typically ask their friend's mom if she will please text me to let me know that they are at her house.

Kids can also practice their texting skills when inviting a friend to play in much the same way as if they were to call. For example:

"Hi Mrs. Jackson. This is Ashley. Can Ella come over and play at our house?"

When they are chatting with friends, texting is a quick, easy and casual way to connect. As your child leans more on texting to interact with friends, provide guidance about times when it would be better to talk rather than text.

When a conversation turns serious or holds the potential to create a misunderstanding, urge your child to phone the friend to discuss the issue. By picking up the phone or meeting to talk in person, your child can address a disagreement and show the other person that he is sensitive to the fact that clarification was needed to avoid or amend hurt feelings.

Although it is becoming more common, breaking up with a significant other via text is still considered a faux pas, right up there with breaking up with someone by ignoring their calls. Encourage your teens to take the higher road and give the other person an explanation either in person or over the phone. It takes courage, but they'll feel better about themselves for it, and it will provide a sense of closure for the other person.

Even with emoticons, sarcasm and jokes don't always play out well via instant message. And of course, whenever there is an urgent or emergency situation, phone is best.

Also, avoid setting your phone to respond with a text message if you receive a phone call that you can't answer. Some phones can be set to automatically respond to a call in text messages like: "Can't talk right now." "I'll call you later." "What's up?"

If callers want to talk via text that's what they would have done in the first place. They know by obvious deduction that you can't talk at that moment since their call went to voice mail. Let your voice mail do the work and then pick up the message and respond when it is convenient for you.

Consider No-Phone Zones

Thanks to smart phones, we have access to each other 24-hours a day through text, instant messaging, email, phone, social media and video apps. As a family, you will want to decide on appropriate boundaries for phone use in your home. Some families guard family and personal time by requiring all phones be turned in at the door or turned over at specific times of the day like dinner time and bedtime, while others have a more informal approach.

While you should do what works best for your family, creating conscious boundaries encourages all of us to focus on the present moment and on the people standing in front of us. The ping of a new message is a strong lure. We have practically programmed ourselves to drop everything, including the person sitting in front of us, to check to see who is calling, texting or emailing.

Consider these factors when trying to explain to your kids why they need to give the person they are with first priority:

- Constantly checking your phone sends a nonverbal message that your friend isn't as important as whoever is hanging out in cyberspace.

- By responding immediately to every single non-urgent communication, we train our contacts that we are available 24-7. Being at everyone's beck and call is exhausting, draining and stressful.

- Many kids hang out together while also on the phone texting with other people. While it might be acceptable in that context, if your child is with extended family members who aren't attached to a device, watching your child bury her head in a phone can come across as disrespectful. Underscore old school communication skills to encourage thoughtful engagement and awareness of their immediate environment.

Be a Role Model

Yes, there's that role modeling reminder again. The truth is our kids watch how we treat our electronic communications when they are with us and use that as a model for what is appropriate. I personally made a conscious decision that if I'm sitting down to dinner with my family, the pings, dings and rings will have to wait 20 to 30 minutes. Dinner is our sacred space. This is the time when I can really connect with my family about the day and relax in the moment.

Naturally, there are exceptions to every rule and we have to maintain some flexibility in a given moment. If you are like most parents, flexibility is your second name! The exception I make to my dinnertime policy is if my husband calls home while he is out of town.

Other times I silence or ignore my phone is during reading time with my kids (again, unless it's their dad calling to say good night), after the lights go out, when I'm with friends (unless it is the sitter), or whenever I need time to concentrate without distraction.

By creating boundaries, I want my children to know that they are my priority in that moment. Plus it's just common courtesy to give the person talking to you your full attention whenever possible. Additionally, I want them to learn that it is perfectly okay to prioritize time for themselves where they choose not to be disturbed.

Common times to set aside phones:

- Family time (dinner, movie or game night).
- While driving.
- At the check out in the grocery store.
- Homework time.
- School or during class.
- Church services.
- Bedtime.

- Social situations like birthday parties, family reunions, weddings, etc., where people are interacting with each other or are expected to be an active observer/listener.
- Funerals.
- During plays, concerts, recitals or other performance events.
- Public places like the coffee shop, movie theater, elevator, library or a restaurant.

If technology overuse seems to be draining you and your family's time or causing a disconnect between each other, try pushing re-set. Agree to a few boundaries and consider setting aside a technology-free weekend once a month or every six weeks. Not sure if you can do it? Try it once to see how you like it and if you feel closer as a family and more energized individually.

Group Texting

Group text conversations are common and are like a party line for friends to talk together. Try to include people in the conversation who know each other. Some people don't feel comfortable responding if they don't know everyone in the audience. Also respect that some people have limited texting plans. When they are pulled into a group text where they receive other people's messages it can cause problems for them by eating into their text quotas. To avoid angering or annoying friends, ask them if you can include them in group texts.

Sometimes conversations can turn ugly if the group starts attacking or making fun of one of the participants in the group or someone who isn't there to defend himself. Encourage your child to follow her personal cues for when a conversation seems to be turning mean-spirited, gossipy or cruel. Does she know everyone who is exchanging messages or are there numbers she doesn't recognize? If she's unable to redirect the conversation, give her some ideas to gracefully exit the group discussion, like "Mom's on my case. Gotta go." Or better yet, encourage her to stand up for the friend who is being maligned.

Since it is relatively simple for a message thread to be forwarded or a screen shot taken and sent to the person who was being discussed, tell kids to assume that anything they say via text is a permanent record of their conversation and will be seen by others. Ask your child if she would be comfortable being a participant in the conversation, especially if she knew her comments would be shared outside the group.

Group Video Chats

Group video chats and hangouts are also popular. Like anything, when used responsibly and with boundaries firmly in place, the technology offers an entertaining and even proactive way for kids to connect with each other, organize clubs/extracurricular activities and help each other with homework—all from the relative safety of home.

Again, remind your kids not to do anything on video that they don't want the entire world to see and to avoid random Internet video chat rooms where people she doesn't know are hanging around. If someone joins the conversation who makes her feel uncomfortable, tell her that she can always use her planned exit strategies to remove herself from the situation.

Here are a few exit strategy ideas you can suggest to your child both for video and text group chats:

> "My time's up. I have a school project/homework coming due that I have to work on."

> "It was good talking to you all! I have to go now."

> "Parents are bugging me. I gotta go."

"I have to get to practice. Talk to you guys later."

"Man, that's just wrong to talk about a girl like that. Hey listen, I gotta go. Catch you guys later."

Above all, common sense and a conscientious approach should keep your child out of trouble and enjoying light-hearted video hangouts with friends.

Chapter 10: Playing in the Gaming Realm

The idea of the 'lone gamer' is really not true anymore. Up to 65 percent of gaming now is social, played either online or in the same room with people we know in real life.

~ Jane McGonigal

Why am I bringing up gaming in a book about communication skills for kids? Many games are prosocial, interactive and create friendship-building opportunities.

When I first started researching gaming for my kids, I was skeptical about how playing video games could offer any redeeming intellectual value. I'd prefer my kids were outside playing with friends, matching wits on board games, riding their bikes, communing with nature, making up imaginary games or doing anything besides sitting in front of a screen that sucks the brain cells out of their developing noggins.

The truth is, unlike television, which is a passive activity, many games require active thinking, concentration, organization, planning, teamwork, creativity and physical movement. Prosocial games nurture empathy-building skills, like caring for a pet or helping neighbors. Enjoyed in moderation, gaming can enhance a child's cognitive and creative abilities.

In addition, the ability to talk about popular games can help kids build social connections with peers. If your child and another child hit it off over a game, schedule a play date when they can play the game together. Many platforms also allow for interactive virtual gaming, which you can set up between your child and his friend.

Safe gaming

Here a few basic tips to help your child play it safe in the world of online gaming:

- Tell him not to "friend" people he doesn't know in real life. Some adult predators pose as children.
- The space invites the possibility for bullying, scammers and hackers. Hang out with your child if he plays online games. If anything dodgy happens, block the person and file a complaint.
- Remind your child not to give out any personal information like his game password, real name, email address or home address.
- Avoid gaming sites that don't feature privacy policies, policies for handling abusive behavior or that aren't well-known.

Gaming is a positive way to connect with your child. He'll probably love the opportunity to
teach you how to play the game. In the meantime, you can monitor what he is playing and with whom he typically plays the games.

If you have trouble distinguishing games that create value versus condone needless violence and criminal behavior, check out CommonSenseMedia.org or LearningWorksforKids.com, which rate video games and apps. Also, brush up on the rating system. For example, games that are rated "M" for mature are technically geared for 18 and up, and often feature blood and extreme violence (including violence against women).

Many teachers use gaming as a way to engage kids in learning math facts, as well as reading and spelling skills. Talk to your child's teacher about games that she recommends for entertaining ways to help your child learn.

You can also use games to help your child learn how to be safe online. NetSmartzkids.org, FBI.gov and PBS Kids Webonauts Academy are just three of the websites I found after a quick search. Participate with your child to answer questions and offer guidance. Who knows, you might learn something new, too.

Chapter 11: Shaping a Smart Shutterbug

Not everybody trusts paintings, but people believe photographs.
~ Ansel Adams

Photo technology offers many opportunities for your kids to explore their creativity in ways like never before. You no longer need expensive camera equipment or photo design software to create professional-looking pictures.

Many kids use their smart phones to take pictures of the food they eat or the silly faces they make to share with friends. They can also use free graphic design tools and programs to add artistic flair and colorful text to their photos.

Toddlers and preschoolers can even get in on the action. Kids' learning technologies typically feature cameras and give them options to "paint" or "draw" with their stylus to enhance the photos.

Other activities your budding shutterbug can try include:

- Making photo mash-ups for their photo-sharing site.
- Creating a collage of photos.
- Building slide shows to share with the family.
- Designing a photo book of their summer vacation.
- Crafting a how-to book or slideshow with pictures.
- Pinning pictures to boards that they organize online.
- Using their talents on their blog and website.
- Volunteering as the photojournalist for the school newspaper or yearbook.
- Documenting club activities.

Another way to grow your child's skills is through specialized spring break or summer camps that teach kids innovative photo technology techniques. Also, introduce your kids to the work of famous photographers and look at professional photographer portfolios available online. Encourage them to experiment with different techniques and emulate the photographers whose work they admire.

Headed to a sibling's baseball or soccer game? Hand your youngster your phone and set it to camera. Ask her to take some photos. I've often found that photos taken by my kids have interesting angles and offer the unique perspective from their vantage point. Who knows, you might find a game-winning gem. And while they are busy entertaining themselves, you can enjoy the game.

Your child may also enjoy creating her own photography blog to share her pictures with friends and family. She can practice captioning and discussing her hobby with others. Another idea is to take a photo a day and do a quick slide show of a vacation or the year.

Photo Precautions

As with anything, impart sage advice to your child to help him learn ethical photo taking and sharing. Photography is simple to upload but without considering possible ramifications could cause heartache for those involved. Taking selfies and documenting mundane day-to-day activities are generally harmless, but it is a good idea to enlighten our children about the rights of others and what photos can communicate.

In our image-centered society, we all love the ability to share photos of our children, our spouses, our pets and other loved ones. With photo sharing, we live vicariously through our friends' vacations. Photos give us a chance to replay those moments that we hold dear to our heart.

People grow more sensitive to photos when they or their children are featured online somewhere, especially if they have a strong sense of privacy or feel powerless to control the circulation of a photo. Whenever in doubt, the best policy is to ask the person.

Once your kids are the ones shooting the pics, teach them to respect other people's boundaries. Explain to your kids that people have different preferences and comfort zones. If one of their friends or another person says they don't want their picture taken, remind your child that is that person's prerogative. Photographing foods and pets is one thing, but party pics or intimate photos can come back to haunt them down the road.

Worth a Thousand Words

The photos we share communicate what we appreciate and value in life. Caught up in the moment, kids often don't consider what a photo might say about them to the world at large. The photos they post may give people the wrong impression about them. Let's say your teen (hypothetically, of course) posts party pix of himself and his friends drinking, smoking and whooping it up for the camera. The background is littered with keg cups and bottles of booze. People are going to make assumptions about your son's character, even if he can normally be found studying, playing soccer and mentoring younger kids at school.

Together educate yourselves on the safety of different photo apps and decide which ones make sense from both an entertainment and security standpoint. Some applications create a false sense of security when it comes to photo sharing, promising that the sender can make the photo disappear within seconds from the other person's phone. These types of apps can encourage impulsive behavior that may cause regret later. With a quick swipe of the phone, a screenshot of the photo can be taken, saved and distributed.

Occasionally flip through your child's photos. Even if you don't find anything concerning, you'll probably get a kick out of them!

It isn't always easy for kids to fathom that their best friend or boy/girlfriend will betray them by distributing a private photo. They are only one argument or break-up away from regretting a photo they entrusted to the person they thought cared about them. The results can be devastating to their reputation. And, pictures can be used as blackmail in abusive relationships.

You have the power to provide educated guidance. Rather than banning technology out of fear (because let's be honest, that's just not realistic), train your child to use it responsibly and keep the lines of communication open to help pave the way for him to come to you in a dicey situation and to make smart choices.

Talk about what is acceptable and what isn't. Discuss problematic situations you see on the news or on popular television shows. Another way to prevent hasty decisions is to go over possible scenarios.

Hypothetical situations to discuss might include:

- You see your friend doing something silly or embarrassing. You snap a picture. What do you do with it?
- Your boyfriend asks you to send a sexy photo. He promises he'll keep it safe and for his eyes only. What do you do or say in response?
- You and a group of your friends are at a party. The party-goers are binge-drinking, smoking, dancing and generally acting loud and crazy. Everyone is taking pictures and posting them immediately to social networks. How do you protect yourself from troublesome photos?

Again, help your child become conscious of personal boundaries and educate each other on the types of apps available and their capabilities. Underscore your family's values and encourage him to trust his intuition about what feels right and what doesn't.

Video Sharing and Vlogging

From the time your children are little, model how you use video sharing. For example, take video of yourself at interesting landmarks while on business trips and send it to your partner or the caretaker you've put in charge to share with your kids. If your kids are still young, you can also record yourself reading one of their favorite books. They'll love listening to you read over and over again. What a wonderful way to connect when you can't be there to tuck your kids in at night!

Look for educational apps, classes or camps that can help your kids safely explore digital media. Besides video, many kid-friendly applications teach kids to make music, write code for video games, and tell stories through video.

Basic video cameras have become inexpensive and are a fun way for kids of all ages to create their own home movies. They can upload the movies onto the computer to edit and share with the family. Someday those short clips will become priceless in your family archives.

They can post their how-to videos on YouTube to help other kids learn to do craft projects or better understand a technology that they've educated themselves on. Review the videos first before your child posts them to ensure that the material doesn't reveal too much about who they are and where they are located.

Your kids might also like to try their hands at a video blog or vlog. Encourage them to explore specific interests and in the beginning, help them set up a private blog just for friends and family. Together, you'll have to do a little research to determine how secure the blog is and if the posts can still be shared despite privacy settings. Nonetheless, your child can learn in a friendly way how to share and develop his talents and voice for an audience.

Vlogs can be a fun, interactive and educational summer project. For example, if your child is interested in nature, together come up with 10 blog post ideas that he can produce. One week he could do a short write up about beetles and then take some video of beetles at the nature center or in the backyard. Or, how about a mash-up video documenting a family vacation in a one second video clip each day or his baseball season set to copyright-free music? The possibilities are endless.

Chapter 12: Cultivating Audio/Video Skills

I'm a great believer that any tool that enhances communication has profound effects in terms of how people can learn from each other, and how they can achieve the kind of freedoms that they're interested in.

~ Bill Gates

Many companies and entrepreneurs use audio and video, like webcasts and podcasts, to reach their audiences. Webcasts are streaming video technology that allow audiences to view video or listen to audio directly on the Internet. They can watch from their computers or on their smart phones. Webcasts are usually created to reach a large audience. Since kids are more likely to podcast and video chat rather than webcast, I will focus on these mediums in this chapter.

Podcasting

Podcasts are audio or video files located on the Internet that are available for download to an mp3 player, phone or computer. Podcasts are often presented in episodes and uploaded to an RSS feed, which audience members can subscribe to. The audiences can be large or small, but podcasts generally appeal to niche audiences.

Learning to podcast offers an innovative and inexpensive avenue for kids to learn communication skills. Plus, it's fun. Who knows, you may have a future broadcaster in your midst who got his start in podcasting.

Wondering how to podcast? Record your audio with a digital recorder, on your iPhone or directly through your computer microphone. Convert the audio to a universal MP3 file. Upload the file to a media host, and then the link is typically added to a blog post.

YouTube features many instructional videos about how to podcast. Have your child listen to a few podcasts to learn how they should sound and look (if there is a visual element). Some schools feature podcasts that kids produce, which might be also be helpful for your child to review.

Why podcast? Podcasts are an on-demand, affordable technology that allows you to speak directly to your audience in a conversational way. Because podcasts are interactive, kids can invite feedback from audience members who download and listen to the mp3 file.

A few of the ways kids can use podcasting include:

- Work individually or as teams to produce school radio shows.
- Interview each other, teachers or visitors to school.
- Present research projects.
- Discuss current events.
- Sing songs
- Play copyright-free music
- Share their writing.
- Tell a story.
- Explain a how-to project.

If your child has a blog, he can post his podcast to the blog for family and friends to download.

Because podcasting is interactive, provide guidance about how to converse with commenters to the post. For example, your child can thank people for comments about the post and answer questions about the material.

> Commenter: "Hey, great podcast. I learned a lot and you did a good job presenting the information."

Your child: "Thank you for listening."

Show them how to tell the difference between real commenters and spam. A spam blocker plug-in can help. For example, people who send spam usually drop a link with no other message, write in a nonsensical way or have strange, unknown IP addresses.

Just as in real life, in audio conferencing platforms, voice projection, articulation and inflection are important to holding the interest of the audience. Some people use slides or other visuals to give their audience something to look at during the podcast. Tutorials about how to include visual images in a podcast can be found online and vary depending on the type of software you own. As your child practices podcasting, encourage her to talk in a confident, conversational tone. And, remind her not to chew gum.

Video Chatting

Chatting through video phone apps with friends is an informal way for kids to grow comfortable talking on camera. Go through your child's contact list with her to decide which of her friends she would like to video chat with. A child can have hundreds of "friends" on social media, but limit video chatting to only people she is close with and trusts. Chances are those are the people she'd prefer to hang out with on video anyway since this is a more intimate type of conversational app.

Even young kids can have fun with video chats, which are especially beneficial for building connections with extended family who live far away. Invite grandparents and aunts and uncles to call. And use video chat if you are traveling away from your youngsters on business.

Here are few activities you can facilitate for your toddler, preschooler and early elementary child to help them engage in video chatting, in between making silly faces at themselves on camera.

- If you are out of town, show your kids your hotel room where you are staying while you talk.
- Chat while pointing out big city landmarks in the background.
- Read books together (make sure you both have a copy of the book to maintain a child's interest).
- Sing songs to each other.
- Share artwork.
- Tell stories. Start a story and have your child finish the story.
- Create comic relief. Some video applications allow you to add facial hair or silly hats that move around as you move on the screen.

Tips for Future Reference

Early exposure can help kids become comfortable using video to communicate. Down the road, your teen or college grad may have to do an interview or conference with a teacher or business leader.

Help your child learn the ins and outs of video conferencing in such formal situations. Background decor, clothing, eye contact, body language and noise can all distract the listener/viewer and detract from the desired message.

Tips for successful formal video chats include:

- Dress to impress. Skip the jammies or rock logo t-shirt with sweatpants. Even if they can't see your sweatpants, you're subconsciously more likely to conduct yourself in a professional manner if you are wearing the right attire to fit the circumstances.

- Clean up. Tidy up the background area. Viewers don't want to see trash, piles of paper or anything too personal in the background.

- Choose a quiet area. Find a space that is free from distracting sounds like barking dogs, loud children, radio music or phones.

- Watch your body language. Sit in a chair that doesn't swivel or move. When we're nervous, it's common to rock, swivel or fidget. Try to appear relaxed, natural and engaged through positive nods, gestures and warm smiles.

- Keep your hands off your face, neck and hair. Because the focus is on your face in video chats, avoid touching your face, massaging the back of your neck (which can look like boredom) or playing with hair or jewelry. Also, don't rub your nose, which is not only unattractive, it signals dishonesty.

- Eye contact. Don't stare blankly at the screen, but do look at the camera when you speak rather than at the other person or at your own image. When you look at the other person, it looks like you are looking down and not at them, which psychologically can make the other person feel like you aren't making good eye contact. Try minimizing your picture to avoid getting distracted and use your peripheral vision to see the person you are speaking with.

Of course, your kids have plenty of time to learn how to video chat professionally, but it never hurts to have a few tips in your back pocket as a reference.

Chapter 13: Choosing Online Groups

A tribe is a group of people connected to one another, connected to a leader, and connected to an idea. For millions of years, human beings have been part of one tribe or another. A group needs only two things to be a tribe: a shared interest and a way to communicate.

~ *Seth Godin*

The Internet is loaded with online groups engaged in various niches and interests. When I was writing my book *Confidently Connected: A Mom's Guide to a Satisfying Social Life*, I encouraged moms to explore online groups to help develop connections. These groups offer support, guidance and a wonderful way to interact with people who share your interests.

As your kids become exposed to online groups, help them differentiate the trash from the treasure. One of the biggest red flags online is groups that allow anonymity. When people are afforded the luxury of cloaked identities, basic civility seems to fly out the window. Hidden behind a screen, some folks seem to think that gives them permission to use derogatory language, make inflammatory comments and say hurtful things to other commenters. This is not an environment that is conducive to positive support and friendship.

Search for groups that make sense for your child's maturity, age level and interests. Perhaps he like writing stories or creating illustrations for stories. There are online groups and websites created for kids that support and nurture this interest. I provide a list of current resources on my website.

Help your child evaluate his priorities when he joins an online group.

- What does he hope to get out of the group?
- Does he want to make friends?

- Do members of the group share a common interest?
- How much personal information does he think is okay to share in the group?
- How does he know who he is really chatting with?

Advise your child to get a feel for the group's dynamics and understand the rules of the group before jumping in, especially if he doesn't know the people involved.

Be an Initiator

Another way to help your child decide what types of groups he values is to help him form his own online group with friends who share common interests. By doing so, he can practice:

- Moderating.
- Creating guidelines and rules for behavior in the group.
- Educating himself on security and protection lock downs.
- Establishing parameters for membership.

As always, tell him to keep boundaries in check by remaining wary of sharing too much personal information, especially if there are members who he doesn't know. Guidelines could include:

- Real picture avatars only—no cartoons, especially if you don't know the person.
- Welcome members and encourage them to introduce themselves.
- Profile member request by Googling them.
- Trust your instincts. If something doesn't seem right about a person, it probably isn't.
- Real names—no pseudonyms.
- No attacking the character of members of the group.
- Differences of opinion encouraged as long as the comments are constructive and presented in a tolerant, respectful way.

Group Blogging

Creating a group blog is another way kids can engage with each other. One person can set up the blog and authorize additional contributors. They can then take turns posting about a particular topic of interest. Help them set up privacy guidelines, create group rules, and establish security measures to defend against spam and hackers.

While your child is busily engaged in numerous school and extracurricular activities, having an online group to discuss photography, gaming, art, music or sports is a positive outlet. In addition it is a good way for kids to network with peers or even reach out to experts who can help them further explore their interests.

Chapter 14: Managing Social Media Gossip and Bullying

I think the hardest part about being a teenager is dealing with other teenagers—the criticism and the ridicule, the gossip and rumors.
~ *Beverly Mitchell*

People will often share tidbits of gossip to start conversations. Discussing news and pop culture is one thing. But, broadcast rumors about those who are closer to home and you enter potentially harmful territory.

For the listener, the desire to hear all of the salacious details, whether those details later prove to be true or false, is practically wired into our DNA. Reality TV, television shows, pop-culture magazines and the tabloids make millions every day thanks to gossip.

Social media can also turn into a hotbed of gossip and rumors, with some sites actually encouraging members to perpetuate negativity. Behind the screen of the web, kids may fall into a false sense of security believing they can spread rumors anonymously without the repercussions of getting caught. Mean jokes, cruel comments and verbal hazing can quickly get out of hand, rippling across networks in seconds.

A recent study found that one in three kids will experience cyber-bullying or be the subject of lies or vicious gossip that threatens to harm their reputation in a very public way.

But the good news is, as parents, we don't need to panic or lose our cool.

- Maintain the same level of open communication that you are already committed to.

- Underscore treating others with courtesy.
- Take practical, proactive measures.
- Provide practical advice and guidance should your child find herself in the midst of contentious circumstances.
- Respect age requirements for social media apps.

Most of all, rely on the empathy skills and boundaries that you have been nurturing in your child from day one to help her manage online conflict—and every online interaction—with dignity and poise.

What to Do if Your Child is Victimized

Social currency is everything to an adolescent and just like for any of us, it is devastating when a child's reputation is called into question or slandered. Adolescents don't necessarily have the maturity or life experience to allow negative situations to roll off their backs. These events are hard enough for adults to endure, with some having to turn to experts to manage the firestorm.

While a child's first impulse might be to retaliate, deny the allegations and go on the defense, these methods can backfire and only create more headaches for the victim.

Take preventative measures.

- Maintain a strong connection with your child to help keep the lines of communication open.
- Discourage her from being part of sites that condone cruel behavior. While these sites may be entertaining, you never know when they can turn on you.
- Be aware that cyber-bullying is illegal in many states. Find out what the laws are in your state and if legal recourse is necessary.
- Maximize privacy and security on all social media sites.
- Have your child's passwords and login information for all accounts.

- Friend your child—not to comment or share, but to quietly keep an eye on what's happening on her accounts.
- If you notice your child making unseemly remarks, pull her aside and talk to her about it.
- Discourage friending people she doesn't know, which can open the door for online harassment.
- Notice who your child decides to follow online to get a sense of the types of individuals she admires. Is there anyone who seems questionable or you haven't heard of before? Talk about it.
- Tell her why it is important to avoid sites that condone anonymity, which seem to lay the groundwork for mean-spirited behavior.
- Help your child build a solid offline support network through extracurricular activities that support her interests and nurture a positive self-image.
- Limit online interaction.
- Remind your child that the golden rule applies online, too. Treat others online as she would like to be treated.
- Teach your child that her online persona is not separate, but rather an extension of her offline identity. Everything she does online is reflective of who she is offline.

If your child is attacked through social media, help her through the fallout.

- Encourage her to confide in you or another trusted adult.
- Advise her to remain cool and reach out to her family and closest friends for support.
- Avoid angry retaliation or defensiveness, which can make the attacks worse.
- If she chooses to speak to the person spreading the rumors, tell her to do it in person and discuss what to say. For example, "I know we don't get along, but I'd like you to stop spreading lies about me."

- If possible, shut down accounts where the attacks are occurring.
- Contact the school's administration if a classmate is the one spreading lies and bullying.
- If a crime has been committed or your child feels threatened, contact local law enforcement.
- Continue engaging in activities that bring her joy and fulfillment.

If Your Child Bullies

If you find out that your child is the one doing the teasing or bullying, tell him it has to stop immediately. He may think it's no big deal. That you are over-reacting. That he's just having a little fun at another person's expense.

Talk about how he would feel if he was the target. Help him see how quickly information travels and how hurtful that information can be. Share examples of cyber-bullying that got out of hand, including how the actions affected both the perpetrator and the victim. And, remind him that every action he makes even in a supposedly anonymous online forum is a reflection of his character and his values.

The best way to judge whether or not an action is right is to mentally check in with questions like, "Will this comment unnecessarily hurt another person? Does this comment feel respectful? When I think about making the comment, do I get a sick feeling inside?"

In the next chapter, let's explore online reputation management some more.

Chapter 15: Online Reputation Management

People are very reluctant to talk about their private lives, but then
you go to the Internet and they're much more open.

~ Paulo Coelho

Our offline and online identities are blurring, which is why it makes sense to consider how one reflects upon the other. In the business of public relations (PR), practitioners strategize about reputation management, controlling the message and the persona of the individuals who represent a brand. As parents, it's a smart move to put on the PR practitioner's hat for a moment as we consider how to help our kids approach social media as an entity that is an extension of their offline life.

Kids are joining their voices and images in droves to video hang outs, photo platforms, chat rooms and other social networks. Can the wrong information land in the wrong hands? Is it possible to say something that they wish they could take back? Sure, which is why I'm providing tips to help your kids learn how to manage their online persona. They can carry these lessons with them into their future lives.

Creating an Authentic Online Profile

As your child sets out to join online networks, discuss a profile that reflects who he is in real-life (without sharing too much personal information like where he lives or what school he attends). Encourage your child to express himself in a natural way, pointing out the importance of showing patience and respect toward others, which can help him through all sorts of social situations. Remind him to be himself while seeking out groups who share his interests. Some kids find more support from friends online than they do at school and the right online forum can be a positive and healthy way to connect.

As he grows more comfortable in social media, remind him to control the impulse to share images or comments that he might regret later with the mantra "If in doubt, I won't post." Try to help him see the long term and evaluate the consequences. Encourage him to ask himself questions like: "Would I want mom and dad to see this photo? How would I feel if the entire school saw it?" Show him examples of mistakes peers have made to help him understand the gravity of risky online behavior.

If your child makes a mistake, discuss how he would approach the problem in real life. Is it appropriate to apologize? Is there something he can do to make up for the mistake?

Friends love to play pranks on one another. To help him avoid becoming vulnerable, encourage him to protect his social media pages by guarding his passwords and logging out if he goes into his account from a friend's computer or in a public setting.

In recent years across the country, some kids have been suspended from school for impersonating school administrators. Online impersonation, even when it starts out as a joke can have serious— even legal—repercussions all around. Discuss potential consequences and help your child understand that cyber-bullying and impersonation is criminal behavior.

Identify Online Role Models

All around us, we see celebrities engaging in outrageous behavior to attract attention and publicity, even if it is negative. People by nature can't help but watch a train wreck. Unfortunately the circus of attention can normalize this sort of behavior to adoring fans and impressionable youngsters.

Look for examples of celebrities or change makers your kids admire, who are careful with how they manage their image. While you should be wary of turning any celebrity into a role model, show your kids how these successful stars carefully control their image and messages to the benefit of their reputation in the long run.

Don't You Like Me?

Your child may feel compelled to post something titillating because she thinks it will gain her the "likes" she craves. "Likes" in our culture equal popularity. "Likes" help us feel validated and make us feel that what we say and who we are matters. Companies also recognize this hunger, and some target teens who are well-liked in social media. The companies that do this understand that teens make purchasing decisions based on what their peers are buying. By targeting the teens with the most likes, they can further promote their products to teens who desire the same gold standard of popularity.

Even young kids are getting in on the action when they ask their parents to post a photo of themselves, while requesting updates on the number of likes the photo received. They are making the connection: "If people like my picture that means they like me."

Try to help your child see that likes don't equal self-worth. An active, fulfilling and rewarding life outside of social media can help ward off some of this insatiable need for social media audience approval.

Reining in Impulsivity

In an impatient culture where instant gratification is more the rule than the exception, controlling impulsive decisions is harder. When emotions are involved in the situation, the stakes—and the risks—rise. Bring this up with your kids.

Ponder the following questions with your kids:

- What would I *want* to do if my best friend betrayed a secret online?
- What *should* I do if she betrays me online?
- What if a girl who likes me sends a provocative picture? Should I share it with my buddies?
- Who should I tell if someone spreads an ugly and damaging rumor about me online?
- What if I see a rumor posted online about a classmate I don't like? Should I share it with my friends for laughs?

Assure them that we all get angry and want to lash out from time to time, but that there are better, healthier ways to manage anger. Share real-life examples, including stories of when you made mistakes and how you dealt with them.

Also, discuss solutions that typically work to calm your child and re-channel negative energy.

Ideas to manage anger or hurt feelings:

- Take 24-48 hours to cool down. Make this a rule.
- Write out frustrations in a journal.
- Talk to a trusted friend, parent or adult.
- Engage in a physical activity like taking a walk, going for a bike ride or a run, dancing or yoga.
- Shoot hoops at the gym.
- Go to a movie.
- Watch a funny TV show.
- Play a video game.
- Listen to music.
- Practice an instrument.
- Draw, sew, paint, build, cook or do anything that is a creative outlet, requires focused attention and takes your mind off the problem for a little while.

Stay Up to Speed

Keeping up with online trends is challenging for parents. Teens and tweens are fickle and switch apps and platforms as frequently as they change their clothes. Websites like CommonSenseMedia.org can help parents stay on top of the trends. Also ask your teens and their friends which apps are the rage at the moment. Listen when they talk to each other in the carpool.

Also, if your child is asking to download a specific app that you don't know much about, ask him to teach you about it. What are the pros? What is the value of the program and what can be learned? What are the cons? What are the risks? Tell him to do his research carefully because you will be following up to see what you can find out on your own. Come up with a set of criteria for downloading apps.

For example:

- What benefits or value does this app add to our life?
- How much does the app cost?
- If there is a fee for the app, is this what he wants to spend his allowance on?
- Does this fit with our family's values?
- Is this safe?
- Are there security issues we need to be aware of?
- What do reviewers say about it?

In this way, your child is taking a conscious approach by educating himself, which will help him learn to critically evaluate future apps that are trending.

Ideally, you will have already established a pattern of communication with your child where he comes to you, seeking guidance about security online and making you aware of problems that he runs into. If you aren't entirely confident that your child is being upfront with you, watchdog software is available to help keep tabs on what kids are up to. You can also set up Google alerts with your kids' names to help you monitor where their name shows up online.

We parents dance a fine line between creating trusted communication with our kids, guiding them through difficult social situations, and keeping tabs on them to prevent a regrettable chain reaction. You have already laid a foundation of tolerance, kindness and respect. Even if they momentarily forget these values in an online or offline interaction—and who doesn't from time to time—you can gently steer them back on track.

The world may be growing faster-paced and more interconnected, but childhood remains a sacred time where our kids should be given the room to learn and grow from their mistakes. We don't typically throw kids into the lake to teach them to swim. In the same vein, with thoughtful, measured steps, you slowly integrate and transition kids into technology. They'll learn the risks like everything else, but revel in the many social and educational benefits available to them like no generation has seen before.

Conclusion

At its core, communication encompasses our ability to relate to others. Communication skills are handed down to children through parental role modeling, conversation, storytelling and woven into the hundreds of interactions kids experience daily both in the real world and the virtual landscape.

As you implement the ideas you have learned in this book, be patient, especially if communicating with others doesn't come easily for your child. Try to avoid labeling your kids as shy or overly friendly. Labels have a way of reinforcing our beliefs about ourselves. A child who identifies herself as shy may start to believe she doesn't have it in her to manage small talk at a party and decides not to bother. A child who is overly friendly may believe his role is to please everyone, and he doesn't give himself permission to create healthy boundaries.

There are people in this world who seem to be naturally skilled communicators. I bet you can think of certain motivational speakers, celebrities, historical figures, authors, a politician or two, or colleagues, who you especially admire for their ability to share their opinions and thoughts in an inspiring way and with integrity.

These traits aren't a result of luck or the right splice of DNA. Many of these folks worked diligently at nurturing these skills.

Some people equate being a good communicator with gregariousness or extroversion. I don't believe this to be true. People can be conscious, thoughtful communicators even if they tend to be introverts. Being an extrovert doesn't necessarily give you an edge. I've known extroverts who struggle with getting their point across in a concise manner, are impatient listeners, or intrude on other people's boundaries.

No matter what their disposition, children can learn to speak with confidence. Will they experience disappointing or hurtful interactions despite their best efforts? Of course. We all do. Comfort them, share your experiences, but most of all, encourage them to keep trying. Reiterate that while it's important to respect other people's boundaries, we can't allow others to dictate how we move through this world.

Communication is an ever-evolving work in progress. By equipping our kids with a strong foundation of healthy communication skills, we will help them grow more self-possessed, feel happier and more secure in their relationships, and to better navigate and thrive in this hyperconnected world in which we live.

Appendix

Stranger Danger? Exploring the Grey Areas

We tell our kids not to talk to people they don't know to help protect them. But, what should they do if they need help and you aren't there?

- With very young kids, avoid confusing the issue and simply tell them not to talk to people they don't know or don't know well.

- Some experts suggest pointing a child toward another mom with children, but others say err on the side of caution as there are female impersonators out there.

- Tell your child to head to a customer service or the store cash registers if he can't find you. Remind him to never leave the building to search for you because you would never leave the building without him.

- When lost, teach your child to share limited information with strangers like his first name, phone number and address. If he doesn't know that, his first name and your first and last name should be sufficient.

- If an adult makes him feel uncomfortable, tell him to yell "Stranger!" and run.

- Create a secret family password. If someone asks your child to go somewhere with him, your child can say that her mom and dad only allow her to go with someone who knows the password. In a real emergency, the person you've put in charge will know the family password.

- Encourage awareness of the environment where ever you go. Discuss options of who he could ask for help if he loses you. What are employees wearing? What do their name tags look like?

- No matter how old you are, the best way to protect yourself is by simply being aware of what is going on around you. Tell your adolescent to put away the phone to be better aware of what is happening in her environment.

- Have an emergency contact list. Make sure your kids know who is on your approved list and who is allowed to take them home.

- Tell them who they are NOT to go with, especially if you have someone untrustworthy in your family.

- Advise your child to never go to a second location with anyone. If a store employee wants your child to go to a back room with him to call you, your child can say "No, I'll stay here. Here is my mom's name and phone number. Could you please call her or call her name over the intercom?"

A Word on Social Media Addiction

Is it possible to become addicted to your online social network? Some professionals argue that social media addiction is a real disorder that creates problems in a person's offline life. Other experts say that the addiction label is tossed around too casually without considering what the term actually means.

Merriam-Webster's dictionary defines addiction as "a strong and harmful need to regularly have something (such as a drug) or do something (such as gamble)." Addiction is usually triggered by emotional stress.

Since adolescents are by nature glued or "addicted" to their social network, it can be difficult for a parent to know for sure if their teen's social media use is normal or not. Instead of hanging out at the mall or sitting on the phone for hours at a time, kids today connect online.

Peer-to-peer contact can become a problem when no boundaries are in place and your child is connected 24-7.

Here are a few signs that your child might need a break from his social media network.

- Texting and social media is getting in the way of quality sleep.
- Chores and other responsibilities aren't being completed.
- Experiences excessive social drama, including bullying and isolation.
- Any time your teen is apart from his phone, he gets stressed about what he might be missing out on.
- Has trouble being present in the moment.
- Becomes a distraction from concentrating on homework.
- Poor performance at school and a drop in grades.

- Prefers online interactions to face-to-face get-togethers.
- Shows signs of depression.

An over-reliance on social media can drive insecurities and affect a teen's mood.

"Teenagers in particular can be very vulnerable to mood changes because they compare themselves to others on social networking sites," says Dr. Gwenn O'Keefe, pediatrician, spokeswoman for the American Academy of Pediatrics, and the author of *CyberSafe: Protecting and Empowering Kids in the Digital World of Texting, Gaming and Social Media.* "There might be kids getting invited to parties or lots of pictures of people doing other things and they start to feel inferior. Often that's a very normal reaction."

Take these steps to prevent social media from dominating your child's life:

- Reinforce boundaries like no-tech zones at home, in the car, at bedtime or around the dinner table. These boundaries can help encourage face-to-face interaction and help your child concentrate on the present moment or tasks at hand.

- Periodically have the family take social media and/or technology vacations. Even a short weekend unplugged can be beneficial, helping family members refocus, enjoy each other's company or simply sit with their own thoughts.

- Treat social media as a privilege and take away access if it is interfering in your child's life. Reintroduce it gradually and monitor carefully.

- Know which social media platforms your child is using. Have access to her social media account passwords.

Randomly check to see what your child is posting and how often.

- Keep computers in a centralized location in your home to help monitor online behavior.

- Don't allow your child to lie about his age. Most popular social media platforms require that users are at least 13 years old in accordance with the "Children's Online Privacy Protection Act." Your child won't take you seriously about digital integrity if you don't consistently insist on following the rules.

- Encourage your child to get involved in extracurricular activities that nurture social skills.

Above all, if your teen is exhibiting abnormal moods that don't improve after two weeks, contact your pediatrician to determine if intervention with a psychologist or therapist is warranted.

Glossary of Terms

Blog - A website used to express personal opinion, experiences, thoughts or photographs. The interactive nature of the site allows the writer to interact with readers who comment on the post. A blogger can limit his or her audience to subscribers only or open it to the public.

Digital citizenship - The established parameters and norms for appropriate, responsible behavior while using technology.

Digital native - A term first coined by author Marc Prensky in 2001 to describe people who were born after the introduction of interactive digital technology.

Hyperconnected - Always connected through electronic devices to other people and the global online world.

Podcast - A digital audio or video file that is posted on a website and available for download or subscription. A podcast is usually part of a themed series and geared toward a niche audience.

Social media - Websites and other forms of online communication that provide a platform for individuals to socialize, share information and connect personally and/or professionally.

Vlog - A video journal posted to a website that is available to viewers to comment on. Like a blog, the platform can be set up to allow for audience interaction with the creator of the journal.

Recommended Resources

The Engaging Child: Raising Children to Speak, Write and Have Relationship Skills Beyond Technology by Maribeth Kuzmeski

Alone Together: Why We Expect More from Technology and Less from Each Other by Sherry Turkle

Generation Text: Raising Well-Adjusted Kids in a World of Instant Everything by Dr. Michael Osit

Playing Smarter in a Digital World by Randy Kulman

CyberSafe: Protecting and Empowering Kids in the Digital World of Texting, Gaming and Social Media by Dr. Gwenn O'Keefe

Additional resources and an evolving list of social media platforms, websites and apps popular among adolescents is available at www.christamelnykhines.com.

Bonus Download!

As a special thank you for purchasing this book, I am offering a free Family Digital Citizenship Contract especially geared to kids who are just beginning to get involved in technology. The contract is available for download at http://www.christamelnykhines.com/free-digital-citizenship-contract/.

Acknowledgments

Thank you to Lara Krupicka, Sue LeBreton and Sara Marchessault for your keen attention to detail and constructive feedback during the editing process. And to Christina Katz for your mentorship and valuable advice each step of the way.

I'd also like to extend a special thank you to each of my first-readers who took the time to review this book. Your thoughtful feedback was critical to helping me deliver a quality product that parents will find both resourceful and practical.

About the Author

In addition to *Happy, Healthy & Hyperconnected: Raise a Thoughtful Communicator in a Digital World*, Christa Melnyk Hines is the author of *Confidently Connected: A Mom's Guide to a Satisfying Social Life*, which helps moms build and sustain a thriving, balanced social life that supports their emotional health.

As a freelance journalist and family communication specialist, Christa enjoys exploring the numerous ways healthy communication creates happier, stronger families.

With *Happy, Healthy & Hyperconnected*, Christa continues to bring to light issues surrounding connection and communication concerning today's families.

Her articles appear monthly in national and regional publications across North America. Christa holds a B.A. in speech communication with a minor in journalism from Texas A&M University and a M.A. in speech communication from the University of Nebraska at Omaha. Research for her master's thesis on the use of social media in public relations expanded her interest in how our modern society communicates both from an individual and a business standpoint. Her article based on her thesis was published in the *Journal of New Communication Research.*

As the mom of two boys, Christa understands the worries that parents have about striking the right balance between the offline and online worlds to help set the tone for a healthy, grounded family. She is encouraged by the many ways in which technology allows families to connect without losing sight of the traditional skills that emphasize thoughtful, considerate communication. She believes that raising kids with a well-rounded set of communication skills will distinguish them in our global high-tech environment as they grow into adulthood and pursue their dreams.

Christa is committed to creating connection in her community both online and off. Among her activities, she facilitates a Facebook group for moms called "Confidently Connected Moms" and coordinates a neighborhood book club. In addition, every month on her blog she champions "Inspiring Moms" who spearhead creative and proactive efforts to build connection among moms in their community.

When she isn't writing, she can be found trying to keep up with her active children and golf-impassioned husband; tempting her family with her various culinary concoctions; lost in a novel; or walking her dogs. She resides with her family in the greater Kansas City metropolitan area.

Connect with Me

Please let me know what you thought of this book and feel free to share your experiences.

Here are a few ways to connect with me:
- Visit me at www.christamelnykhines.com.
- Sign up for my free monthly e-newsletter, which features inspiring tips, ideas and the latest news and resources on building and sustaining connection.
- Connect by email at christahines13@gmail.com.
- Follow me on Twitter @ChristaHines1.
- Find me on Pinterest: Cnhines4
- Join the conversation in my Facebook group for moms at "Confidently Connected Moms." Although the group is closed to protect the privacy of members, it isn't exclusive. Simply submit a request to join, and I'll add you to the group.

www.ingramcontent.com/pod-product-compliance
Lightning Source LLC
Chambersburg PA
CBHW071454070426
42452CB00039B/1353